CRUDE OIL DISTILLATION

UNLOCKING THE SECRETS OF HYDROCARBON SEPARATION

Copyright@2023

Trysta Brown

TABLE OF CONTENT

CHAPTER I. INTRODUCTION

DEFINITION AND IMPORTANCE OF CRUDE OIL DISTILLATION

Crude oil distillation is a primary process in petroleum refining that involves the separation of crude oil into various fractions based on their boiling points. It is a crucial step in the production of various valuable products, such as gasoline, diesel, jet fuel, and heating oil, among others. Crude oil is a complex mixture of hydrocarbons, containing different molecular weights and boiling points. The purpose of crude oil distillation is to separate these hydrocarbons into different fractions, each with a specific range of boiling points. This separation is achieved by heating the crude oil and allowing it to vaporize, then condensing the vapors at different temperatures to collect the desired fractions.

IMPORTANCE OF CRUDE OIL DISTILLATION

crude oil distillation is a vital process in petroleum refining, enabling the separation of crude oil into various valuable fractions. Its importance lies in the production of essential fuels and feedstock for downstream processes, meeting energy demands, driving economic growth, and addressing environmental considerations.

1. Production of valuable products

Crude oil distillation is vital for the production of a wide range of valuable petroleum products. These products play a significant role in various industries and sectors of the economy. Gasoline, diesel, jet fuel, and heating oil are some of the major products obtained through distillation. These fuels are essential for transportation, heating, and powering various machinery and equipment.

2. Meeting energy demands

The distillation process enables the extraction of energy-dense hydrocarbon fuels from crude oil. These fuels serve as a primary source of energy worldwide, meeting the high demands for transportation, industrial processes, and electricity generation. Crude oil distillation plays a crucial role in ensuring a reliable and consistent supply of energy to support economic activities.

3. Feedstock for downstream processes

The fractions obtained from crude oil distillation serve as feedstock for various downstream processes in petroleum refining. These fractions undergo further processing, such as cracking, reforming, and treating, to produce additional valuable products. For instance, naphtha, a fraction obtained during distillation, can be further processed to produce petrochemicals used in the production of plastics, synthetic fibers, and other chemical products.

4. Economic impact

The petroleum refining industry, which relies heavily on crude oil distillation, has a significant economic impact. It creates employment opportunities, generates revenue through exports, and contributes to the overall economic growth of producing regions and countries. Moreover, the availability and affordability of petroleum products obtained through distillation contribute to the smooth functioning of various sectors, including transportation, manufacturing, and agriculture.

5. Environmental considerations

While crude oil distillation primarily focuses on separating valuable fractions, it also enables the removal of impurities and contaminants from crude oil. This process contributes to reducing the environmental impact associated with the use of crude oil. Additionally, advancements in refining technologies and environmental regulations have led to the development of cleaner and more efficient processes, mitigating the environmental footprint of the refining industry.

The petroleum refining process involves the conversion of crude oil into various valuable products through a series of physical and chemical processes. Now is a brief outline of the petroleum refining process:

1. Pre-Treatment:

- Desalting: Crude oil is treated to remove salts and other impurities.
- Distillation: The crude oil is heated and partially distilled to remove volatile components and separate it into different fractions based on their boiling points.

2. Conversion Processes:

- **Cracking:** Heavy hydrocarbon molecules are broken down into lighter molecules through thermal cracking (high-temperature heating) or catalytic cracking (using catalysts).

- **Reforming:** Molecules are restructured to increase their octane rating and improve gasoline quality.

3. Treatment Processes:

- **Hydrotreating:** The fractions are treated with hydrogen under high pressure and temperature to remove sulfur, nitrogen, and other impurities.

- **Isomerization:** Straight-chain hydrocarbons are converted into branched-chain isomers, improving the properties of gasoline.

- **Alkylation:** Combining olefins and isobutane to produce high-octane gasoline components.

4. Separation and Fractionation:

- **Distillation Towers:** The remaining fractions from pre-treatment and conversion processes are further separated in distillation towers based on their boiling points. Lighter fractions rise to the top, while heavier fractions collect at the bottom.

5. Product Treatment and Blending:

- Various treatments and additives are employed to improve the quality of the separated fractions. These processes include removing impurities, adjusting octane ratings, and achieving specific product specifications.

- **Blending:** Different fractions and additives are combined to create customized gasoline, diesel, jet fuel, lubricants, and other petroleum products.

6. Final Product Distribution:

The refined products are stored in tanks and distributed through pipelines, trucks, or ships to fulfill market demands. Throughout the refining process, environmental considerations and regulations are addressed to minimize emissions, manage waste, and ensure the safe handling of hazardous materials. It's important to note that this overview provides a simplified outline, and the actual petroleum refining process can be much more complex and involve additional steps and processes depending on the specific refinery configuration, feedstock properties, and product requirements.

CHAPTER II. CRUDE OIL COMPOSITION

OVERVIEW OF CRUDE OIL

Crude oil is a naturally occurring, unrefined fossil fuel that is found beneath the Earth's surface. It is formed over millions of years from the remains of plants and microscopic organisms that lived in ancient oceans and lakes. Crude oil is a complex mixture of hydrocarbons, which are organic compounds composed primarily of carbon and hydrogen atoms.

1. Composition:

- **Hydrocarbons:** The main components of crude oil are hydrocarbons, which can be classified into three main groups: paraffins (alkanes), naphthenes (cycloalkanes), and aromatics (aromatic hydrocarbons). The proportions of these hydrocarbon groups vary, giving different types of crude oil distinct characteristics.

- **Sulfur and Nitrogen Compounds:** Crude oil also contains varying amounts of sulfur and nitrogen compounds, which can have environmental and processing implications. Lower sulfur content is desirable as it reduces pollution and the need for extensive refining.

2. Types of Crude Oil:

Crude oil can be classified into different types or grades based on various factors, including geographic origin, density, sulfur content, and viscosity. Some common crude oil types include Brent crude, West Texas Intermediate (WTI) crude, Dubai crude, and heavy sour crudes.

- **Light Crude Oil:** Light crude oil has a lower density, lower viscosity, and higher proportion of smaller hydrocarbon molecules, making it easier to refine into gasoline and other light products.
- **Heavy Crude Oil:** Heavy crude oil has a higher density, higher viscosity, and higher

proportion of larger and heavier hydrocarbon molecules, requiring more complex refining processes to convert it into valuable products.

3. Reserves and Production:

Crude oil reserves are the estimated quantities of oil that can be economically recovered using existing technology. The largest reserves are found in countries such as Saudi Arabia, Venezuela, Canada, Iran, and Iraq. Crude oil production involves the extraction of oil from underground reservoirs using drilling techniques. The production of crude oil is a vital industry, with major producers including the United States, Saudi Arabia, Russia, China, and Canada.

4. Uses and Applications:

Crude oil is primarily used as a feedstock in the production of various energy sources and products. The most significant use is the production of fuels, including gasoline, diesel, jet fuel, and heating oil, which power transportation and provide energy for heating and industrial processes. Petrochemicals: Crude oil is also a key source of feedstock for the petrochemical industry, which produces a wide range of chemicals used in plastics, synthetic fibers, pharmaceuticals, fertilizers, and other products.

5. Global Importance: Crude oil plays a vital role in the global economy, serving as a major source of energy for industries and transportation. Oil prices and availability have a significant impact on global markets, geopolitical dynamics, and the economies of both producing and consuming countries. It's important to note that the properties and composition of crude oil can vary significantly depending on the source and the geological conditions under which it was formed. These variations have implications for refining processes, product yields, and environmental considerations.

TYPES OF CRUDE OIL

Crude oil can be classified into different types or grades based on various factors, including geographic origin, density, sulfur content, and viscosity. Here are some of the well-known varieties of crude oil:

1. Brent Crude: Brent crude is a light, sweet crude oil sourced from the North Sea. It is considered a benchmark for pricing other crude oils, and its relatively low sulfur content makes it easier to refine into high-quality fuels.

2. West Texas Intermediate (WTI): WTI is a high-quality, light, sweet crude oil produced in the United States. It is the benchmark for pricing crude oil in the Americas, and its low sulfur content and low viscosity make it ideal for producing gasoline and other high-value products.

3. Dubai Crude: Dubai crude is a medium sour crude oil sourced from Dubai, United Arab Emirates. It has a higher sulfur content than Brent crude or WTI, making it more difficult and expensive to refine into usable products.

4. Heavy Sour Crude: Heavy sour crude oil is a type of crude oil that has a high density, high viscosity, and high sulfur content. It is more challenging to refine than lighter crude oils and typically requires more complex refining processes. Some examples of heavy sour crude include Venezuela's Orinoco Heavy, Saudi Arabia's Arabian Heavy, and Canada's Western Canadian Select.

5. Light Sweet Crude: Light sweet crude is a type of crude oil that is low in sulfur and has a low density, making it easier to refine into high-value products such as gasoline, diesel, and jet fuel.

Examples include Nigeria's Bonny Light and Angola's Girassol.

6. Sour Crude: Sour crude is a type of crude oil that contains a high concentration of sulfur compounds, which can corrode refinery equipment and produce higher levels of pollution. Examples of sour crude include Mexico's Maya crude and Canada's Syncrude.

COMPOSITION OF CRUDE OIL

Crude oil is a complex mixture of hydrocarbons, organic compounds composed primarily of carbon and hydrogen atoms. Its composition can vary depending on the source and geological conditions under which it was formed.

1. Hydrocarbons:

- **Paraffins (alkanes):** These are straight-chain or branched hydrocarbons. They have the general formula C_nH_{2n+2} and are characterized by their relatively stable and unreactive nature.

- **Naphthenes (cycloalkanes):** Naphthenes are cyclic hydrocarbons with the general formula CnH2n. They have a ring structure and are typically more reactive than paraffins.

- **Aromatics (aromatic hydrocarbons):** Aromatics have a ring structure with alternating double and single bonds. The most common aromatic hydrocarbon is benzene (C6H6), and it is characterized by its distinctive odor.

2. Sulfur Compounds:

Crude oil often contains sulfur compounds, such as hydrogen sulfide (H2S) and various organic sulfur compounds. The sulfur content in crude oil can range from trace amounts to several percent. High sulfur content is known as "sour" crude oil, while low sulfur content is referred to as "sweet" crude oil.

3. Nitrogen Compounds:

Crude oil may contain various nitrogen compounds, such as pyridine and quinoline. Nitrogen compounds in crude oil can contribute to the formation of pollutants during combustion and can be challenging to remove during refining.

4. Oxygen Compounds:

Crude oil can contain trace amounts of oxygen compounds, such as phenols and organic acids. However, the oxygen content in crude oil is generally low compared to other elements.

5. Trace Elements:

Crude oil may contain trace amounts of other elements, including metals such as nickel, vanadium, and iron. These trace elements can have implications for refining processes and can contribute to the formation of catalyst poisons and other operational challenges. It's important to note that the specific composition of crude oil can vary widely depending on its source, with different crude oil types having distinct compositions and ratios of hydrocarbons and other compounds. The composition of crude oil influences its properties, such as density, viscosity, and flammability, and it also affects the refining processes required to convert crude oil into various products.

CHAPTER III. PRE-DISTILLATION PROCESSES

DESALTING

Desalting is an important process in the petroleum refining industry that is carried out to remove salts and other water-soluble impurities from crude oil. Crude oil extracted from underground reservoirs often contains dissolved salts, water, and other contaminants. These impurities can cause various operational and product quality issues during the refining process. Desalting is performed to mitigate these problems and improve the quality of the crude oil before it enters the refining units. Desalting is typically the first step in the crude oil refining process and is performed at the refinery site. It helps protect downstream equipment, such as pumps, pipelines, and heat exchangers, from corrosion and fouling. Additionally, desalting enhances the efficiency of subsequent refining processes and ensures the production of high-quality petroleum products.

1. Purpose:

- **Removal of salts:** The primary objective of desalting is to remove inorganic salts, such as chlorides and sulfates, which can cause corrosion in refinery equipment and catalyst deactivation.

- **Removal of water:** Desalting also helps in eliminating water, which can interfere with refining processes, reduce efficiency, and impact product quality.

2. Desalting Methods:

- **Chemical Desalting:** This method involves mixing the crude oil with a dilution water and an emulsion-breaking chemical, such as demulsifiers or surfactants. The mixture is vigorously agitated to break the emulsion and facilitate the separation of the water and salt contaminants from the crude oil.

- **Electrostatic Desalting:** In this method, high-voltage electrical fields are used to separate the water and salt particles from the crude oil. The crude oil is passed through an electrostatic field, which causes the charged water and salt particles to coalesce and settle out of the oil.

3. Desalter Equipment:

- **Desalter:** A desalter is a vessel or unit specifically designed for the desalting process. It typically consists of a mixing section where the crude oil, dilution water, and emulsion-breaking chemicals are combined and agitated. It is followed by a settling section where the water and salt contaminants separate from the oil.

- **Heating Elements:** Heat may be applied to the crude oil in the desalter to reduce its viscosity and improve the separation efficiency.

4. Process Operation:

- Crude oil and dilution water are introduced into the desalter unit.
- Emulsion-breaking chemicals are added to aid in the separation of water and salts.
- The mixture is agitated to promote the coalescence of water and salt particles.
- The settled water and salt contaminants are removed from the bottom of the desalter.
- The desalted crude oil, with reduced salt and water content, is sent to the next refining units.

HEATING AND PREHEATING

Heating and preheating are essential processes in the petroleum refining industry that involve the application of heat to various streams and feedstock. These processes serve several purposes, including reducing viscosity, enhancing reaction rates, improving separation efficiency, and optimizing energy utilization. Here's an overview of heating and preheating in the context of petroleum refining:

1. Heating:

Heating involves raising the temperature of a substance, typically a feedstock or a process stream, to a desired level. Heat can be applied through various methods, including direct firing, steam, hot gases, or heat transfer fluids.

Purpose of heating

- **Reducing Viscosity:** Heating crude oil or heavy hydrocarbon fractions reduces their viscosity, making them easier to handle and process.

- **Facilitating Separation:** Heating can aid in the separation of different components in a mixture, such as distillation of crude oil into various fractions based on their boiling points.

- **Promoting Reactions:** Heating can increase the rate of chemical reactions, such as hydrocarbon cracking or reforming, by providing the necessary activation energy.

- **Enhancing Heat Transfer:** Heating improves heat transfer efficiency, allowing for more effective utilization of heat in the refining processes.

- **Improving Pumpability:** Heating certain feedstocks, such as heavy residues or bitumen, improves their flow characteristics and facilitates transportation through pipelines or other conveyance systems.

2. Preheating:

Preheating involves raising the temperature of a feedstock or a process stream before it enters a specific unit or process within the refinery. Preheating is typically achieved by utilizing heat exchangers, where the heat from a hot stream is transferred to a colder stream.

Purpose of pre heating:

- **Energy Optimization:** Preheating helps to optimize energy usage in the refinery by utilizing waste heat or excess heat from one process to preheat another stream, reducing the overall energy consumption.

- **Process Efficiency:** Preheating the feedstock or process streams before they enter various units improves the efficiency of subsequent processes, such as distillation, catalytic cracking, or hydro treating.

- **Minimizing Fouling:** Preheating can help prevent fouling or deposits in equipment by ensuring that the feedstock or streams are at the required temperature, preventing the precipitation of solids or the condensation of heavy components.

Preheating is commonly employed in different refinery units, including crude oil heaters, furnaces, heat exchangers, and process heaters.

The specific temperatures and heating requirements vary depending on the process, feedstock properties, and desired product specifications. Overall, heating and preheating play crucial roles in the petroleum refining industry, enabling efficient processing, improving product quality, and maximizing energy utilization.

CRUDE OIL STORAGE AND HANDLING

Crude oil storage and handling are important aspects of the petroleum industry, involving the storage, transportation, and management of crude oil from production sites to refineries or other destinations.

1. Storage Facilities:

- **Tank Farms:** Crude oil is stored in large tank farms, which consist of multiple storage tanks. These tanks are typically made of steel and can vary in size and capacity, ranging from a few thousand barrels to millions of barrels.

- **Floating Storage:** In some cases, crude oil may be stored in floating storage facilities, such as floating storage and offloading (FSO) or floating production storage and offloading (FPSO) vessels. These vessels are moored offshore and can store and offload crude oil directly from production platforms or tanker ships.

Underground Storage: Crude oil can also be stored underground in natural rock formations, such as salt caverns or depleted oil and gas reservoirs. Underground storage provides additional security and helps to maintain stable temperatures for the stored oil.

2. Handling and Transportation:

- **Loading and Unloading:** Crude oil is transferred between storage tanks and transportation vessels, such as tankers or pipelines, through loading and unloading operations. These operations involve the use of pumps, valves, and specialized equipment to transfer the crude oil safely and efficiently.

- **Pipeline Transportation:** Pipelines are commonly used for the transportation of crude oil over long distances. Crude oil is pumped into pipelines and transported to refineries or other storage facilities. Pipeline networks are a critical infrastructure for the oil industry, providing a reliable and cost-effective means of transportation.

- **Tanker Transportation:** Crude oil can be transported via tanker ships, which are specifically designed for carrying large quantities of oil across oceans and waterways. Tanker vessels come in different sizes, from small coastal tankers to massive supertankers known as very large crude carriers (VLCCs) or ultra-large crude carriers (ULCCs).

3. Safety and Environmental Considerations:

- **Spill Prevention and Response:** The handling of crude oil involves strict safety measures to prevent spills and leaks. Containment systems, monitoring equipment, and spill response plans are in place to mitigate the environmental impact in case of an incident.

- **Vapor Emission Control:** Crude oil emits volatile organic compounds (VOCs), which can contribute to air pollution. Vapor recovery systems and controls are employed to capture and control VOC emissions during storage and handling operations.

- **Environmental Protection:** Proper management of storage facilities includes measures to prevent soil and groundwater contamination, such as lining tanks and implementing spill containment systems.

4. Quality Control and Sampling:

- **Quality Monitoring:** Crude oil quality is an essential consideration. Regular sampling and analysis of the stored crude oil are conducted to ensure it meets the required specifications and to detect any impurities or contaminants that may affect refining processes or product quality.

- **Blending:** In some cases, crude oil from different sources or grades is blended to achieve the desired properties and specifications for refining or market requirements. Blending operations are carried out to optimize the value and consistency of the crude oil.

CHAPTER IV. DISTILLATION TOWER

OVERVIEW OF DISTILLATION TOWER

A distillation tower, also known as a distillation column or fractionation tower, is a key component in the petroleum refining industry. It is a tall vertical structure that facilitates the separation of various components of crude oil or other feedstock based on their boiling points.

1. Purpose:

- **Fractionation:** The main purpose of a distillation tower is to fractionate the crude oil or feedstock into different components or fractions based on their boiling points. Each fraction obtained from the tower has a specific range of boiling points and contains hydrocarbons with similar properties.

- **Product Separation:** The distillation tower separates the crude oil into lighter fractions, such as gases, gasoline, diesel, and jet fuel, as well as heavier fractions, such as kerosene, fuel oil, and residual fuel. This separation allows for the production of various refined petroleum products.

2. Tower Structure:

- **Height and Diameter:** Distillation towers are tall structures, typically ranging from tens to hundreds of feet in height. The diameter of the tower decreases as it ascends to maintain the necessary pressure and temperature conditions for separation.

- **Trays or Packing:** Inside the tower, there are a series of horizontal trays or packing materials. These provide surfaces for liquid-vapor contact and create multiple stages for the separation process. Trays are perforated plates with holes or slots, while packing consists of materials like metal or ceramic structured packing or random packing materials.

- **Reflux System:** The distillation tower incorporates a reflux system that allows the condensed liquid from the upper part of the tower (overhead) to flow back down as reflux. The reflux helps improve separation efficiency and control the temperature profile within the tower

3. Operation:

- **Feed Entry:** The feedstock, such as crude oil, enters the distillation tower at a specific level, known as the feed entry point. The feed is introduced either at the bottom of the tower or at an intermediate stage, depending on the desired separation requirements.

- **Separation Stages:** As the feed rises through the tower, it is heated, and various components with different boiling points begin to vaporize. The vapor rises through the trays or packing, while the liquid flows downward. The liquid and vapor phases come into contact, allowing for the transfer of heat and mass.

- **Fractionation:** The components with lower boiling points rise to the upper sections of the tower, while the components with higher boiling points tend to condense and flow downward. This separation is achieved due to the temperature gradient within the tower, with higher temperatures at the bottom and lower temperatures at the top.

- **Product Collection:** The separated fractions are collected at different levels in the tower. The lighter components are collected from the upper part (overhead) of the tower, while the heavier fractions are withdrawn from lower stages or the bottom of the tower.

4. Side Streams and Reflux:

- **Side Streams:** Distillation towers may have side streams extracted at specific levels to remove impurities or to obtain intermediate products with desired properties.

- **Reflux:** A portion of the condensed liquid from the overhead product is returned to the top of the tower as reflux. The reflux helps to maintain temperature control and optimize separation efficiency.

Distillation towers are a vital component in the petroleum refining process, allowing for the separation and production of different refined petroleum products. The specific design, configuration, and operation of a distillation tower depend on the feedstock properties, desired product specifications, and the refining objectives of a particular refinery.

COMPONENTS OF THE DISTILLATION TOWER

The distillation tower, also known as a distillation column or fractionation tower, is a tall vertical structure that separates crude oil into different components or fractions based on their boiling points. The following are the main components of a distillation tower:

FURNACE

A furnace is a key component in the petroleum refining industry that is used for heating various process streams, feedstock, or other fluids. It is an enclosed structure where combustion of fuel takes place, generating heat that is transferred to the material being heated. Furnaces are crucial in providing the necessary thermal energy for several refining processes. Here is a summary of furnaces in the context of petroleum refining:

1. Purpose:

- **Heat Generation:** The primary purpose of a furnace is to generate heat by burning fuel, typically natural gas, fuel oil, or refinery gas. The heat produced is used for various refining processes that require elevated temperatures, such as cracking, reforming, distillation, and hydrotreating.

- **Temperature Control:** Furnaces play a crucial role in controlling and maintaining precise temperature profiles required for specific reactions or processes within the refining units.

2. Furnace Types:

- **Fired Furnaces:** These are the most common type of furnaces used in refineries. They have a burner system that directly ignites and combusts the fuel inside the furnace chamber. The hot gases produced by the combustion process transfer heat to the material being heated.

- **Radiant Tubes or Radiant Coil Furnaces:** These furnaces use radiant tubes or radiant coils to transfer heat to the material. The fuel is burned outside the radiant tubes or coils, and the hot combustion gases pass over the tubes, transferring heat through radiation.

- **Fluidized Bed Furnaces:** In fluidized bed furnaces, a bed of solid particles is fluidized by blowing air or another gas through the particles. The fuel is burned within the fluidized bed, and the intense mixing and contact between the fuel and particles result in efficient heat transfer.

3. Furnace Components:

- **Combustion Chamber:** The combustion chamber is the central part of the furnace where the fuel is burned. It provides a controlled environment for the combustion process and contains the necessary equipment, such as burners, igniters, and flame detectors.

- **Heat Transfer Surfaces:** Furnaces have heat transfer surfaces, such as radiant tubes, coils, or refractory-lined walls, which absorb and transfer heat from the combustion gases to the material being heated.

- **Fuel and Air Supply System:** Furnaces have fuel supply systems that deliver the required fuel, and air supply systems that provide the necessary combustion air for the fuel to burn efficiently.

- **Flue Gas System:** The flue gas system collects and removes the combustion gases, typically through a stack or chimney. It may also include pollution control equipment to reduce emissions and comply with environmental regulations.

4. Safety and Control Systems:

- Safety Measures: Furnaces incorporate safety systems, such as flame detectors, fuel trip systems, and pressure relief devices, to ensure safe and reliable operation.

- Temperature and Process Control: Furnaces are equipped with instrumentation and control systems to monitor and control the temperature, fuel-air ratio, and other parameters, ensuring optimal operation and process control.

Furnaces in petroleum refining are designed and operated with a focus on energy efficiency, environmental compliance, and operational safety. The specific design and configuration of a furnace depend on the process requirements, fuel availability, desired temperature ranges, and regulatory standards.

PRE-FLASH ZONE

The pre-flash zone is a section within a crude oil distillation column or a distillation tower where a partial vaporization of the crude oil feed occurs before entering the main distillation section. It is located at the lower part of the distillation tower, typically just above the bottom of the column.

1. Purpose:

Vaporization: The primary purpose of the pre-flash zone is to partially vaporize the crude oil feed. By introducing heat to the feed in this section, a portion of the more volatile components of the crude oil vaporize, forming a mixture of vapor and liquid. This partial vaporization helps to reduce the load on the main distillation section and improves the overall efficiency of the distillation process.

2. Operation:

- **Heat Addition:** In the pre-flash zone, heat is added to the crude oil feed through various means, such as heat exchangers or direct steam injection. This heating raises the temperature of the feed, causing the more volatile components to vaporize.

- **Partial Vaporization:** As the heated crude oil enters the pre-flash zone, the lighter hydrocarbon components, such as gases and light naphtha, vaporize due to their lower boiling points. These vapors rise upward while the heavier liquid components, such as heavy naphtha, kerosene, and gas oil, continue to flow downward.

- **Separation:** The vapors from the pre-flash zone, which contain the more volatile components, are separated from the remaining liquid. They are typically routed to the top of the distillation tower for further separation and collection as overhead products. The remaining liquid, still containing the heavier components, continues its flow downward into the main distillation section of the tower.

3. Advantages:

- **Load Reduction:** The pre-flash zone helps to reduce the load on the main distillation section by separating and removing a portion of the more volatile components early in the process. This allows for better separation efficiency and reduces the risk of flooding or other operational issues in the main distillation section.

- **Improved Product Quality:** By partially vaporizing the crude oil feed, the pre-flash zone helps to separate and remove undesirable components, such as light gases or light ends, which may have lower value or cause operational challenges downstream. This results in improved product quality for the desired fractions.

- **Energy Efficiency:** The pre-flash zone can contribute to energy efficiency by allowing for the recovery of heat from the overhead vapors. The heat recovered can be utilized for various purposes within the refining process, such as preheating other process streams or generating steam.

DISTILLATION TRAYS OR PACKING

Distillation trays and packing are key components within a distillation tower or column. They provide surfaces for the contact between liquid and vapor phases, facilitating the separation and fractionation of different components based on their boiling points. Here is a summary of distillation trays and packing:

1. Distillation Trays:

Purpose: Trays are perforated plates or decks that create multiple stages within the distillation column. They provide surfaces for liquid-vapor contact and allow for the transfer of mass and heat between the phases.

Types of Trays: There are various types of trays used in distillation towers, including sieve trays, valve trays, bubble cap trays, and structured trays. Each type has a specific design and mechanism to enhance vapor-liquid contact and separation efficiency.

- **Sieve Trays:** Sieve trays consist of a perforated plate with holes or slots, allowing vapor to pass through while holding back liquid. The perforations help distribute the vapor across the tray and provide points for vapor-liquid mixing.

- **Valve Trays**: Valve trays have movable valves or caps that open and close as vapor passes through. The valves help create turbulence and promote intimate contact between the vapor and liquid.

- **Bubble Cap Trays:** Bubble cap trays have individual caps or domes over each tray hole. As vapor rises, it lifts the caps, creating bubbles and increasing vapor-liquid contact.

- **Structured Trays:** Structured trays utilize structured packing elements, such as corrugated or structured sheets, to provide a large surface area for vapor-liquid interaction. They can improve separation efficiency and reduce pressure drop compared to conventional trays.

2. Packing Materials:

Purpose: Packing materials are alternative options to trays and are used to create a large interfacial area between the liquid and vapor phases. They facilitate vapor-liquid contact, allowing for efficient mass transfer and separation.

- **Types of Packing:** There are different types of packing materials, including random packing and structured packing.
  Random Packing: Random packing consists of small, irregularly shaped pieces

of materials, such as ceramic, metal, or plastic, that are randomly filled in the column. Examples of random packing include Raschig rings, Pall rings, and Berl saddles. Random packing creates a tortuous path for vapor and liquid flow, promoting contact and separation.

- **Structured Packing:** Structured packing comprises arranged elements or sheets with specific geometric shapes, such as corrugated or honeycomb patterns. Structured packing provides a well-defined and uniform surface area for vapor-liquid interaction, improving separation efficiency and reducing pressure drop.

REFLUX SYSTEM

The reflux system is a vital component of a distillation process, particularly in a distillation tower or column. It plays a crucial role in enhancing separation efficiency and maintaining optimal temperature profiles within the tower. The reflux system involves the circulation and controlled reintroduction of condensed liquid back into the distillation column.

1. Purpose:

- **Enhancing Separation Efficiency:** The primary purpose of the reflux system is to improve the separation efficiency of the distillation process. By returning a portion of the condensed liquid from the overhead vapor as reflux, the system increases the liquid holdup and provides additional opportunities for the separation of lighter components from the vapor phase.

- **Temperature Control:** The reflux system also helps to control and maintain the temperature profile within the distillation tower. By reintroducing the condensed liquid at specific stages, it helps regulate the temperature at those points and ensures optimal conditions for the separation of desired fractions.

2. Operation:

- **Collection of Condensed Liquid:** The overhead vapors produced in the distillation tower are condensed using a condenser located at the top of the tower. The condensed liquid, which contains a range of lighter components, is collected and routed to a reflux drum or vessel.

- **Reintroduction as Reflux:** From the reflux drum, a controlled portion of the condensed liquid is withdrawn and reintroduced at various stages of the distillation tower. The reflux is typically introduced at a level above the feed entry point, allowing it to flow downward countercurrent to the rising vapors.

- **Vapor-Liquid Contact:** As the reflux flows downward, it comes into contact with the rising vapors in the tower. This contact facilitates the transfer of heat and mass between the liquid and vapor phases, promoting further separation of components based on their boiling points.

- **Separation of Light and Heavy Fractions:** The reflux system helps separate the lighter components, which are preferentially condensed and returned as reflux, from the heavier components that continue to rise in the tower.

3. Reflux Ratio:

Reflux ratio refers to the ratio of the amount of liquid returned as reflux to the amount of liquid withdrawn as product from the distillation tower. It is an important parameter in determining the separation efficiency and product quality. The reflux ratio is typically controlled and adjusted to optimize the separation performance. Higher reflux ratios can enhance separation efficiency but may require more energy for reboiling and condensing, while lower reflux ratios may result in reduced separation efficiency.

4. Benefits:

Improved Separation: The reflux system improves separation by increasing the liquid holdup and promoting additional vapor-liquid contact within the tower. This leads to better separation of components with different boiling points.

Temperature Control: By reintroducing the condensed liquid at specific stages, the reflux system helps maintain temperature control within the distillation tower. This allows for optimal separation conditions and prevents excessive temperature variations.

Product Purity: The reflux system contributes to the production of purer fractions by ensuring that lighter components are effectively condensed and separated from the desired product streams.

5. Overhead system

The overhead system is an important component in a distillation process, particularly in a distillation tower or column. It is responsible for collecting and handling the vapor and liquid streams that exit the top of the tower. The overhead system plays a crucial role in separating and collecting the lighter components of the feedstock.

1. Purpose:

- **Vapor Collection:** The primary purpose of the overhead system is to collect the vapor that rises to the top of the distillation tower. This vapor contains the lighter components or overhead product fractions that have lower boiling points and higher vapor pressures.

- **Liquid Handling:** The overhead system also collects the condensed liquid formed when the vapor is cooled and condensed. This liquid may contain some desirable components that require further processing or separation.

2. Operation:

- **Condensation:** The vapor that exits the top of the distillation tower is routed to a condenser. The condenser cools the vapor, causing it to condense into a liquid. The condensed liquid is then collected in an overhead receiver or drum.

- **Phase Separation:** The overhead receiver allows for the separation of the condensed liquid and any remaining vapor. The liquid fraction is typically withdrawn from the bottom of the overhead receiver and is often routed to further processing or storage as an intermediate product.

- **Vapor Handling:** The remaining vapor or gases in the overhead receiver are typically removed from the system using a vacuum system, a compressor, or other methods. These gases may be treated or processed separately, depending on their composition and value.

3. Product Separation:

- The overhead system separates the lighter components from the crude oil or feedstock, enabling the production of various valuable products. These products may include light gases such as methane, ethane, propane, butanes, and light naphtha fractions. Depending on the desired product specifications and refinery configuration, the overhead system may also include additional separation equipment, such as knock-out drums, reflux drums, or demisters, to further refine and separate the desired product streams.

4. Safety Considerations:

- The overhead system typically includes safety devices such as pressure relief valves to ensure the safe operation of the distillation tower. These devices protect against overpressure or potential hazards caused by excessive pressure in the system.

- The overhead system may also incorporate measures to prevent the release of volatile or flammable components into the environment. These measures can include the use of vapor recovery systems, flaring, or other emission control mechanisms.

- The design and configuration of the overhead system depend on several factors, including the feedstock properties, desired product specifications, and refinery operating conditions. By efficiently collecting and handling the overhead product fractions, refineries can separate and collect valuable components while ensuring the safe and reliable operation of the distillation process.

PROCESS FLOW WITHIN THE DISTILLATION TOWER

The process flow within a distillation tower, also known as a fractionation column, involves the movement of vapor and liquid phases as they undergo separation based on their different boiling points. Here is a broad overview of the process flow within a distillation tower:

1. Introduction of Feedstock:

The crude oil or other feedstock enters the distillation tower at a specific level, often referred to as the feed tray or feed stage. The feedstock is typically preheated before entering the tower to facilitate vaporization.

2. Rising Vapor Phase:

As heat is applied to the bottom of the distillation tower, the feedstock undergoes vaporization. The vapor phase, which contains the lighter and more volatile components of the feedstock, rises up the tower.

3. Liquid Downflow:

Simultaneously, the liquid phase, which consists of the heavier components that have not vaporized, flows downward through the distillation tower. This liquid downflow is essential for separation as it allows for contact with the rising vapor.

4. Vapor-Liquid Contact:

The vapor and liquid phases come into contact on each tray or packing section of the distillation tower. This contact facilitates the transfer of heat and mass between the phases and allows for separation based on differences in boiling points.

5. Tray or Packing Efficiency:

The trays or packing within the tower are designed to maximize vapor-liquid contact and separation efficiency. Trays provide surfaces for the liquid to flow across and for vapor to pass through, while packing materials increase the interfacial area between the phases.

6. Condensation and Reflux:

As the rising vapor reaches the cooler regions of the distillation tower, it begins to condense. The condensed liquid, known as reflux, flows downward and is collected in the lower sections of the tower. The reflux is used to improve separation efficiency by providing additional liquid holdup and enhancing vapor-liquid contact. It is reintroduced at specific levels in the tower to facilitate further separation.

7. Withdrawal of Product Streams:

Different product streams are withdrawn from the distillation tower at various stages or trays. These product streams correspond to specific fractions with different boiling points and compositions. Lighter components with lower boiling points, such as gases and light liquids, are typically withdrawn from the upper sections of the tower as overhead products. Heavier components are withdrawn from lower sections as bottom products.

8. Temperature and Pressure Profile:

The temperature and pressure within the distillation tower vary along its height. The temperature typically decreases from the bottom to the top of the tower, while the pressure may vary based on specific design requirements.

9. Reboiler and Condenser:

The distillation tower is equipped with a reboiler at the bottom and a condenser at the top. The reboiler provides heat to the tower, promoting vaporization, while the condenser cools the vapor, causing it to condense into liquid form.

CHAPTER V. DISTILLATION PROCESS

ATMOSPHERIC DISTILLATION

Atmospheric distillation, also known as atmospheric crude oil distillation, is the primary process in petroleum refining that separates crude oil into its various fractions based on their boiling points. It is typically the first step in the refining process and serves as the basis for further refining operations. Here is a summary of atmospheric distillation:

1. Purpose:

Separation of Crude Oil: The main purpose of atmospheric distillation is to separate crude oil into different fractions, including gases, gasoline, diesel, kerosene, and heavier components such as fuel oils and residues. This separation is achieved by utilizing the differences in boiling points of the various hydrocarbon compounds present in the crude oil.

2. Process Description:

- **Heating:** The crude oil is preheated and then introduced into a distillation column, also known as a distillation tower. The preheating helps reduce the viscosity of the crude oil and improves its flow characteristics.

- **Vaporization:** In the distillation column, the crude oil is heated further in a furnace or reboiler at the bottom of the column. The heat causes the lighter hydrocarbon compounds to vaporize, forming a mixture of vapor and liquid.

- **Fractionation:** As the vapor rises up the distillation column, it encounters trays or packing materials that provide surfaces for vapor-liquid contact. The contact promotes the separation of the vapor mixture into different fractions based on their boiling points.

- **Condensation:** As the vapor moves up the column, it begins to cool down. At specific levels within the column, known as tray or packing sections, the vapor condenses back into liquid form. This condensed liquid is collected on trays or flows through the packing material.

- **Overhead Products:** The top section of the distillation column, called the overhead system, collects the lightest components of the crude oil that have low boiling points. These overhead products typically include gases such as methane, ethane, propane, and butane, as well as light naphtha.

- **Side Draw and Bottom Products:** Intermediate fractions, such as gasoline, kerosene, diesel, and heavier components like fuel oil, are withdrawn from side draws at specific tray levels. The heaviest components, such as residual or atmospheric residue, are collected as bottom products.

3. Temperature and Pressure:

The temperature and pressure within the distillation column vary along its height. The temperature decreases as you move up the column, while the pressure typically decreases slightly. The specific temperature and pressure profiles are carefully controlled to achieve the desired separation and product specifications.

4. Reflux:

To enhance separation efficiency, a portion of the condensed liquid from the overhead condenser is typically returned to the top of the distillation column as reflux. The reflux increases the liquid holdup and promotes better separation of the desired fractions.

5. Product Quality and Specifications:

The efficiency of the atmospheric distillation process determines the quality and composition of the various fractions obtained. These fractions are further processed in subsequent refining steps to meet specific product specifications, such as octane rating for gasoline or sulfur content for diesel.

Atmospheric distillation is a critical step in the petroleum refining process, enabling the initial separation of crude oil into its different components. The separated fractions serve as feedstock for subsequent refining processes, allowing for the production of a wide range of refined petroleum products.

VACUUM DISTILLATION

Vacuum distillation is a process used in petroleum refining to separate and refine the heavier fractions of crude oil that have higher boiling points and are not easily separated through atmospheric distillation. It is an extension of the distillation process that operates under reduced pressure, allowing for the distillation of high-boiling point components at lower temperatures to prevent their thermal decomposition.

1. Purpose:

- **Vacuum Residue Separation:** The primary purpose of vacuum distillation is to separate the residual or high-boiling components, known as vacuum residue or bottoms, from the atmospheric distillation process. These components have extremely high boiling points and are not easily vaporized under atmospheric pressure.

2. Process Description:

- **Reduced Pressure:** In vacuum distillation, the pressure within the distillation column is significantly reduced compared to atmospheric distillation. The reduced pressure lowers the boiling points of the heavy components, enabling their separation at lower temperatures.

- **Heating:** The vacuum residue is typically heated in a vacuum distillation furnace or reboiler. The heating reduces the viscosity of the feedstock, enhances vaporization, and facilitates the separation of the high-boiling components.

- **Vaporization:** As the vacuum residue is heated, it undergoes vaporization, forming a mixture of vapor and liquid. The vapor rises through the vacuum distillation column.

- **Fractionation:** Similar to atmospheric distillation, the vacuum distillation column contains trays or packing materials that facilitate vapor-liquid contact and separation. The rising vapor comes into contact with the descending liquid, promoting separation based on boiling point differences.

- **Condensation:** As the vapor rises in the column, it begins to cool down. At specific tray levels, the vapor condenses back into liquid form. The condensed liquid is collected on trays or flows through the packing material.

- **Overhead Products:** The lighter components that have lower boiling points and are vaporized under vacuum conditions are collected as overhead products. These products typically include vacuum gas oil (VGO), which serves as a feedstock for further refining processes.

- **Bottom Products:** The heaviest components, including the vacuum residue, are collected as bottom products. The vacuum residue can be further processed or blended with other feedstock for various applications, such as fuel oil production.

3. Temperature and Pressure:

Vacuum distillation operates at significantly reduced pressures compared to atmospheric distillation. The pressure within the vacuum distillation column can range from a few millibars to a few torrs, depending on the specific requirements and design of the process. The temperature is controlled to prevent the thermal cracking or decomposition of the high-boiling components.

4. Reflux and Vacuum System:

Similar to atmospheric distillation, the vacuum distillation process may utilize reflux to enhance separation efficiency. Reflux is a portion of the condensed liquid returned to the top of the column to provide additional liquid holdup and improve the separation of desired fractions. The vacuum system, including vacuum pumps and condensers, is employed to maintain the required reduced pressure within the vacuum distillation column. It helps to maintain the desired temperature profile and enable the separation of high-boiling components. Vacuum distillation is a crucial process in petroleum refining as it allows for the separation and refining of heavy crude oil fractions that cannot be effectively separated through atmospheric distillation alone. By operating under reduced pressure, vacuum distillation enables the production of valuable products such as vacuum gas oil, as well as the extraction of further heavy components from the vacuum residue for other applications.

PURPOSE AND BENEFITS OF VACUMM DISTILLATION

The purpose of vacuum distillation in petroleum refining is to separate and refine the heavier fractions of crude oil that have high boiling points and are not easily separated through atmospheric distillation. The vacuum distillation process offers several benefits and serves important purposes in the refining industry.

1. Separation of High-Boiling Components:
The primary purpose of vacuum distillation is to separate and recover high-boiling components, such as heavy fuel oils and residual fractions, from crude oil. These components have boiling points well above the temperature range of atmospheric distillation and require reduced pressure conditions for efficient separation.

2. Increased Yield of Valuable Products:

Vacuum distillation allows for the extraction of additional valuable products from the crude oil that would otherwise remain in the residue or be lost during atmospheric distillation. By separating and refining the high-boiling components, vacuum distillation increases the yield of valuable products such as vacuum gas oil (VGO) and heavy fuel oils, which can be further processed or used as feedstock in various downstream processes.

3. Improvement in Product Quality:

Vacuum distillation helps improve the quality of the refined products obtained from crude oil. By removing the heavier and more complex components, vacuum distillation can reduce the sulfur content, viscosity, and other undesirable properties of the feedstock, resulting in higher-quality products.

4. Thermal Stress Reduction:

Vacuum distillation operates at reduced pressures, which lowers the boiling points of the heavy fractions being processed. This reduces the thermal stress on the crude oil components and helps minimize thermal cracking or decomposition of the feedstock. By operating at lower temperatures compared to atmospheric distillation, vacuum distillation can preserve the integrity of the heavy fractions and prevent excessive thermal degradation.

5. Flexibility in Feedstock Processing:

Vacuum distillation provides flexibility in processing different types of feedstock, including heavy and high-viscosity crude oils. It enables the refining of a wider range of crude oil sources, including heavy and sour crudes that require specialized processing to extract valuable products. Vacuum distillation can handle feedstock with higher levels of impurities, such as sulfur and metals, which are typically found in the heavier fractions of crude oil.

6. Integration with Downstream Processes:
Vacuum distillation plays a crucial role in the integration of downstream refining processes. The products obtained from vacuum distillation, such as VGO, can serve as feedstock for various secondary refining units, including catalytic cracking, hydrocracking, and coking units. The integration of vacuum distillation with these processes enables the production of a broader range of refined petroleum products.

PRODUCTS OBTAINED FROM VACUUM DISTILLATION

Vacuum distillation is a crucial step in petroleum refining that allows for the separation and extraction of various valuable products from the heavier fractions of crude oil. The specific products obtained from vacuum distillation can vary depending on factors such as the composition of the crude oil, operating conditions, and desired product specifications. There are some common products obtained from vacuum distillation:

1. Vacuum Gas Oil (VGO):

Vacuum gas oil is a valuable product obtained from vacuum distillation. It has a lower boiling point range compared to the residue and contains a mixture of medium to heavy hydrocarbon compounds. VGO serves as a feedstock for secondary refining processes such as catalytic cracking, hydrocracking, or coking units. These processes further convert VGO into lighter, more valuable products such as gasoline, diesel, and jet fuel.

2. Heavy Fuel Oil:

Vacuum distillation produces heavy fuel oils, also known as residual fuel oils or bunker fuels. These fuels have high viscosity and are typically used in industrial boilers, power plants, and marine applications as a source of heat or energy. Heavy fuel oils have a higher carbon content and lower volatility compared to lighter fuels like gasoline or diesel.

3. Asphalt:

Another product derived from the vacuum distillation process is asphalt, which is commonly used in road construction and roofing materials. Asphalt is obtained from the heaviest fraction of the vacuum residue. It is a dense, highly viscous, and sticky material that solidifies at ambient temperatures.

4. Vacuum Residue:

The vacuum residue, also known as atmospheric bottoms or vacuum bottoms, is the heaviest fraction obtained from vacuum distillation. It consists of high-boiling hydrocarbons, as well as impurities such as sulfur, metals, and carbon residue. The vacuum residue can undergo further processing in secondary refining units, such as coking or visbreaking, to produce additional valuable products like petroleum coke or reduced viscosities of residual fuels.

It's important to note that the specific product yields and qualities can vary depending on the crude oil feedstock and the configuration and operating conditions of the vacuum distillation unit. Refineries may also implement additional refining processes downstream of vacuum distillation to further upgrade and convert the obtained products into more valuable and marketable forms.

CHAPTER VI. PRODUCT FRACTIONATION

OVERVIEW OF PRODUCT FRACTIONATION

Product fractionation, also known as product separation or product distillation, is a crucial step in the refining process after crude oil distillation. It involves further separation and purification of the intermediate products obtained from the distillation process to produce various final products with specific characteristics and specifications.

1. Purpose:

The purpose of product fractionation is to separate the intermediate products obtained from the distillation process into different fractions based on their boiling points and desired product specifications. This allows for the production of final products with specific qualities and properties required for different applications.

2. Fractionation Units:

Product fractionation is typically carried out in specialized fractionation units or towers. These towers are similar in concept to the distillation towers used in crude oil distillation but are designed to separate specific product streams based on their boiling points and desired product specifications.

3. Operating Principles:

The fractionation process operates on the principle of fractional distillation, utilizing the difference in boiling points of the components within the intermediate product stream. The product stream is introduced into the fractionation tower at a specific location, and heat is applied to the tower, typically through reboilers or furnaces. The tower is equipped with trays or packing materials that provide surface area for vapor-liquid contact and separation. As the product stream is heated, the components with lower boiling points vaporize and rise up the tower, while the components with higher boiling points remain as liquid and collect at the bottom. The rising vapor undergoes condensation as it encounters cooler temperatures in the upper sections of the tower. The condensed liquid is collected on trays or flows down through the packing material, and the process is repeated on each tray or section of the tower.

4. Separation of Fractions:

Product fractionation allows for the separation of various fractions based on their boiling points and desired product specifications. The tower is typically equipped with multiple trays or sections, each designed to separate a specific fraction. The exact number and configuration of trays depend on the desired product slate and the specifications of the final products. The separation of fractions is achieved by adjusting the temperature and pressure conditions within the tower. Lower-boiling components will condense and collect at higher levels in the tower, while higher-boiling components will condense and collect at lower levels.

5. Product Streams:

The product fractionation process results in the production of different product streams or fractions, each with specific qualities and specifications. The specific products obtained depend on the refinery configuration, feedstock characteristics, and market demands. Common product fractions include gasoline, kerosene, diesel, jet fuel, light and heavy gas oils, and various fuel oil grades.

6. Further Treatment:

The fractions obtained from product fractionation may undergo further treatment, such as hydro treating, catalytic cracking, or blending, to meet specific product specifications, remove impurities, or enhance product quality. These additional processes aim to improve the performance and characteristics of the final products. Product fractionation is a critical step in the refining process that enables the production of a wide range of final products with specific qualities and specifications. By separating the intermediate products obtained from crude oil distillation, product fractionation ensures that the final products meet the desired standards and are suitable for various applications in the transportation, industrial, and commercial sectors.

TYPES OF PRODUCT FRACTIONS

Product fractionation in petroleum refining can yield a variety of product fractions, each with its own specific properties and applications. The specific types of product fractions obtained depend on the refinery configuration, feedstock characteristics, and market demands. Here are some common types of product fractions obtained through fractionation:

LIGHT DISTILLATES

Light distillates are a group of product fractions obtained through the distillation process in petroleum refining. These fractions consist of hydrocarbon compounds with relatively low boiling points. They are characterized by their volatility, light molecular weight, and high flammability. Now are some examples of light distillates:

1. Liquefied Petroleum Gas (LPG):

LPG is a mixture of propane and butane, which are highly volatile and gaseous at ambient temperatures. LPG is commonly used as a fuel for heating, cooking, and various industrial applications. It is stored and transported in pressurized containers.

2. Gasoline:

Gasoline is a widely used light distillate fuel primarily used as a transportation fuel for automobiles and motorcycles. It is volatile, highly flammable, and characterized by its ability to combust readily in spark-ignition engines. Gasoline is produced in different grades with varying octane ratings to meet specific performance requirements.

3. Naphtha:

Naphtha is a light distillate fraction with a boiling point range between gasoline and kerosene. It is a versatile feedstock for the petrochemical industry and is used in the production of various products, including solvents, plastics, synthetic fibers, and detergents.

4. Jet Fuel:

Jet fuel, also known as aviation turbine fuel (ATF), is a light distillate used as a fuel for aircraft engines. It is highly refined to meet stringent specifications regarding flash point, freeze point, and thermal stability. Jet fuel provides the necessary energy for aviation while maintaining safety and performance standards.

5. Kerosene:

Kerosene is a middle distillate fraction with a higher boiling point than gasoline and lighter than diesel fuel. It is commonly used as a fuel for heating, lighting, and certain industrial applications. Kerosene is also used as jet fuel in certain aircraft and as a solvent in various industries.

Light distillates play a crucial role in various sectors, including transportation, residential and commercial heating, and industrial applications. They are characterized by their high energy content, volatility, and suitability for applications that require readily combustible fuels. The production and composition of light distillates can vary depending on the refinery configuration, crude oil source, and specific market requirements.

MIDDLE DISTILLATES

Middle distillates are a group of product fractions obtained through the distillation process in petroleum refining. These fractions have higher boiling points compared to light distillates but lower boiling points than heavy distillates. They are characterized by their moderate volatility and are widely used in various applications. Below are a few instances of middle distillates:

1. Diesel Fuel:

Diesel fuel is a commonly known middle distillate used as a fuel for diesel engines. It has a higher energy density compared to gasoline and is widely used in transportation, including trucks, buses, trains, and some automobiles. Diesel fuel is also used in off-road equipment and as a heating fuel in certain applications.

2. Heating Oil:

Heating oil, also referred to as fuel oil, is a middle distillate used primarily for residential, commercial, and industrial heating purposes. It is similar to diesel fuel but often has slightly different specifications and is dyed for tax purposes. Heating oil is commonly used in furnaces, boilers, and other heating systems.

3. Kerosene:

Kerosene, mentioned earlier as a light distillate, can also fall into the middle distillate category. It is used as a fuel for heating, lighting, and specific industrial applications. Kerosene is commonly used in jet engines as aviation turbine fuel (ATF) and is known as Jet A or Jet A-1 fuel.

4. Gas Oil:

Gas oil is a general term that encompasses a range of middle distillates with varying specifications and applications. It is used as a feedstock for various refining processes, such as catalytic cracking and hydrocracking, to produce lighter products like gasoline and diesel. Gas oil may also be used as a fuel in certain industrial applications. Middle distillates, particularly diesel fuel, play a significant role in transportation, industrial processes, and heating applications. They provide a balance between energy density, stability, and ease of use. The production and composition of middle distillates can vary depending on the specific refinery configuration, crude oil feedstock, and regional market requirements.

HEAVY DISTILLATES

Heavy distillates, also known as residual fuels or heavy fuel oils, are product fractions obtained through the distillation process in petroleum refining. These fractions have higher boiling points compared to light and middle distillates and are characterized by their viscosity and higher molecular weight. Below are some instances of heavy distillates:

1. Fuel Oil:

Fuel oil is a generic term used to describe various heavy distillate fractions that are used as fuel for industrial processes and power generation. These fuels have high viscosity and typically require heating for efficient combustion. Different grades of fuel oil, such as fuel oil No. 6 (commonly known as Bunker C), are used in applications such as marine vessels, power plants, and large industrial boilers.

2. Residual Fuel Oil:

Residual fuel oil, also known as residual fuel or bunker fuel, is a heavy distillate obtained from the bottom of the distillation tower or from vacuum distillation. It is used primarily in marine engines, large power plants, and industrial boilers. Residual fuel oil has a high energy content and is less refined compared to lighter distillates.

3. Asphalt and Bitumen:

Asphalt and bitumen are heavy distillate fractions that have a solid-like consistency at ambient temperatures but become flowable when heated. These materials are commonly used in road construction, roofing, and waterproofing applications due to their durability and adhesive properties. Asphalt and bitumen are produced from the heaviest fractions obtained in the refining process.

4. Petroleum Coke:

Petroleum coke, or pet coke, is a solid carbonaceous material produced as a byproduct of the refining process. It is formed from the residue left behind after the distillation of crude oil or other petroleum feedstock. Petroleum coke has various industrial applications, such as fuel for cement kilns and power plants, as well as a raw material in the production of electrodes for the aluminum and steel industries.

Heavy distillates are primarily used in industrial applications that require high energy content and are less concerned with combustion characteristics or volatility. These fractions are often less refined and may contain higher levels of impurities compared to lighter distillates. The production and availability of heavy distillates can vary depending on the refinery configuration, crude oil source, and market demand for specific products.

USES AND APPLICATIONS OF EACH PRODUCT FRACTION

Below are the common uses and applications of each product fraction obtained through the petroleum refining process:

1. Liquefied Petroleum Gas (LPG):

LPG is used as a fuel for heating, cooking, and certain industrial applications. It is commonly used in households, restaurants, and commercial establishments for cooking and heating purposes. LPG is also used as a fuel in camping stoves, portable heaters, and as a propellant in aerosol products.

2. Gasoline:

Gasoline is primarily used as a fuel for automobiles, motorcycles, and small engines. It is the main transportation fuel and provides the energy required for internal combustion engines. Gasoline is also used as a solvent in certain industrial processes and as a component in the production of various chemicals.

3. Naphtha:

Naphtha is an important fecdstock for thc petrochemical industry. It is used as a raw material for producing a wide range of products, including plastics, synthetic fibers, solvents, and detergents.

Naphtha is a vital component in the production of ethylene and propylene, which are building blocks for many plastics and chemical intermediates.

4. Jet Fuel (Kerosene):

Jet fuel, also known as aviation turbine fuel (ATF), is used as a fuel for aircraft engines. It provides the necessary energy for aviation while meeting strict specifications regarding flash point, freeze point, and thermal stability. Jet fuel is designed to burn efficiently in jet engines and is a critical component for the aviation industry.

5. Diesel Fuel:

Diesel fuel is widely used as a fuel for diesel engines, which power various applications such as trucks, buses, trains, ships, and generators. Diesel fuel provides efficient energy conversion and is known for its high energy density. It is commonly used in transportation, industrial processes, and as a heating fuel in certain applications.

6. Heating Oil:

Heating oil, also known as fuel oil, is used for residential, commercial, and industrial heating purposes. It is commonly used in furnaces, boilers, and space heaters to provide warmth and heat. Heating oil is particularly important in regions where natural gas or electricity is not readily available for heating.

7. Gas Oil:

Gas oil is used as a feedstock for further refining processes such as catalytic cracking and hydrocracking. It can be further processed to produce lighter products such as gasoline and diesel. Gas oil may also be used as a fuel in certain industrial applications or as a blending component for specific fuel formulations.

8. Residual Fuel Oil:

Residual fuel oil, or bunker fuel, is used as a fuel in large-scale applications such as marine vessels, power plants, and industrial boilers. It is commonly used in applications that require high energy output, where the fuel's viscosity and combustion characteristics are less critical.

9. Asphalt and Bitumen:

Asphalt and bitumen are used primarily in road construction for surfacing and paving roads. They provide durability, weather resistance, and a smooth driving surface. Asphalt is also used in roofing materials, waterproofing applications, and in the production of various industrial compounds. Each product fraction serves specific purposes and meets different industry needs. The applications mentioned above are general and can vary based on regional requirements, market demands, and specific product specifications. Refineries often adjust their production ratios and processes to meet the demand for different product fractions in the market.

CHAPTER VII. ADDITIONAL PROCESSES

CRACKING

Cracking is a process used in petroleum refining to break down larger hydrocarbon molecules into smaller, more valuable molecules. It involves the breaking of carbon-carbon bonds in hydrocarbon molecules, which can be achieved through various methods. Cracking is primarily used to convert heavy and less valuable hydrocarbon fractions into lighter, more desirable products, such as gasoline and diesel fuel. Cracking is a vital process in the petroleum refining industry, allowing refineries to convert a wide range of feedstock into more valuable products. It helps meet the demand for transportation fuels, petrochemicals, and other refined products while improving the overall efficiency and profitability of the refining process. Cracking processes have several benefits and play a crucial role in the petroleum refining industry:

1. Production of Lighter Fractions: Cracking enables the conversion of heavier hydrocarbon fractions, such as gas oils or residues, into lighter and more valuable products like gasoline, diesel, and jet fuel. This helps meet the demand for transportation fuels and high-value petrochemical feedstock.

2. Increased Product Yields: By breaking down larger molecules, cracking processes increase the yield of desired products from a given amount of crude oil. This maximizes the utilization of the feedstock and enhances the profitability of the refining operation.

3. Removal of Impurities: Cracking can also help remove impurities present in heavy feedstock. The process can break down sulfur and nitrogen compounds, reducing the sulfur and nitrogen content in the resulting products.

4. Enhanced Refinery Flexibility: Cracking processes provide refineries with the flexibility to adjust product yields based on market demands and changing regulations. By converting heavy fractions into lighter products, refineries can optimize their production to meet specific requirements and maximize profitability.

The two main types of cracking processes are thermal cracking and catalytic cracking.

THERMAL CRACKING

Thermal cracking, also known as pyrolysis or steam cracking, is a process used in petroleum refining to break down larger hydrocarbon molecules into smaller, more valuable molecules. It involves the application of heat without the presence of a catalyst. Thermal cracking is typically carried out at high temperatures (typically 750-900°C) and sometimes with the addition of steam. In thermal cracking, the feedstock, which is usually heavy hydrocarbon fractions such as gas oils or residues, is heated to the desired cracking temperature. The high temperatures cause the hydrocarbon molecules to break apart, forming smaller fragments. This process is driven by the breaking of carbon-carbon bonds within the molecules. Thermal cracking can result in a wide range of products, depending on the specific feedstock and operating conditions. The main products obtained from thermal cracking are:

1. Gasoline: Thermal cracking can produce a significant amount of gasoline, which is a valuable transportation fuel. The cracked gasoline is rich in light hydrocarbons, such as olefins and aromatics, which have high octane ratings.

2. Light Gas Oils: Light gas oils, which are intermediate distillate fractions, are also produced through thermal cracking. These fractions can be further processed or blended to meet specific product requirements.

3. Olefins: Thermal cracking is an important source of olefins, including ethylene and propylene. Olefins are key building blocks for the production of various petrochemicals, such as plastics, synthetic fibers, and solvents.

4. Light Gases: The cracking process generates a significant amount of light gases, including methane, ethane, propane, and butanes. These gases can be used as fuel or as feedstock for other processes.

Thermal cracking is often employed to convert heavier feedstock, such as vacuum gas oils or residues, into lighter and more valuable products. It is an important process for maximizing the yield of valuable products from crude oil and enhancing the overall efficiency of petroleum refining operations. However, thermal cracking has some limitations. It tends to produce a broad range of products, which may require further processing and separation to obtain specific product fractions. Additionally, the process can generate coke as a byproduct, which can deposit on the reactor walls and reduce efficiency. To mitigate these issues, catalysts are used in catalytic cracking processes, which offer better control over product distribution and reduce coke formation.

Overall, thermal cracking is a crucial process in the petroleum refining industry, allowing for the conversion of heavy hydrocarbon feedstock into lighter, more valuable products like gasoline, light gas oils, and olefins. It plays a significant role in meeting the demand for transportation fuels and petrochemical feedstock.

CATALYTIC CRACKING

Catalytic cracking, also known as cat cracking or fluid catalytic cracking (FCC), is a process used in petroleum refining to convert heavy hydrocarbon fractions into lighter, more valuable products. It involves the use of a catalyst to facilitate the cracking reactions at lower temperatures and pressures compared to thermal cracking. Catalytic cracking holds significant importance within the refining industry as one of its pivotal processes.

The catalytic cracking process typically involves the following steps:

1. Feedstock Preheating: The heavy hydrocarbon feedstock, such as gas oils or vacuum gas oils, is preheated to the desired temperature before entering the cracking reactor. The feedstock is vaporized, which allows it to mix more easily with the catalyst.

2. Contact with Catalyst: The preheated feedstock is mixed with a powdered catalyst, usually composed of zeolites or other acidic materials. The catalyst acts as a catalyst, promoting the cracking reactions. The catalyst and feedstock mixture is then introduced into the cracking reactor.

3. Cracking Reactions: Inside the cracking reactor, the feedstock and catalyst mixture undergoes cracking reactions. The high temperature and the presence of the catalyst cause the larger hydrocarbon molecules to break into smaller fragments. Carbon-carbon bonds are broken, resulting in the formation of lighter hydrocarbon molecules.

4. Product Separation: After the cracking reactions, the mixture of cracked products and catalyst passes into a separation system. The separation process involves the separation of cracked hydrocarbons from the catalyst particles. The cracked hydrocarbons are sent to further processing units for separation and purification, while the spent catalyst is regenerated or replaced with fresh catalyst.

Catalytic cracking offers several advantages over thermal cracking:

1. Selectivity: The use of a catalyst allows for better control over the cracking reactions and product distribution. It enables the selective production of desired products, such as gasoline, light gas oils, and olefins, while minimizing the formation of unwanted byproducts.

2. Lower Operating Conditions: Catalytic cracking operates at lower temperatures and pressures compared to thermal cracking. This reduces energy consumption and equipment requirements, leading to cost savings.

3. Catalyst Regeneration: The catalyst used in catalytic cracking can be regenerated after each cycle. The spent catalyst is regenerated by removing the deposited coke and restoring its catalytic activity. This allows for the continuous use of the catalyst, resulting in longer catalyst life and reduced catalyst consumption.

4. Higher Product Yields: Catalytic cracking typically yields higher quantities of valuable products, such as gasoline and light olefins, compared to thermal cracking. This maximizes the utilization of the feedstock and improves the overall efficiency of the refining process.

The products obtained from catalytic cracking, depending on the specific feedstock and operating conditions, include gasoline, light gas oils, olefins, and light gases. These products are essential for meeting the demand for transportation fuels, petrochemical feedstock, and other refined products.

REFORMING

Reforming is a catalytic process used in petroleum refining to convert low-octane naphtha into high-octane gasoline and other valuable products. It involves the restructuring and reformation of hydrocarbon molecules to increase their octane rating and improve their quality. The reforming process typically utilizes a catalyst, commonly platinum or platinum-rhenium, to facilitate the desired chemical reactions. The main reactions that occur during reforming include:

1. Isomerization: Isomerization involves the rearrangement of hydrocarbon molecules to produce isomers. In reforming, straight-chain hydrocarbons are converted into branched-chain isomers, which have higher octane ratings. Isomerization enhances the performance of gasoline by improving its anti-knock properties.

2. Dehydrogenation: Dehydrogenation is the removal of hydrogen atoms from hydrocarbon molecules. During reforming, dehydrogenation reactions occur, converting cyclohexanes and naphthenes into aromatic hydrocarbons. Aromatics have higher octane ratings and contribute to the overall quality of the gasoline product.

3. Hydrocracking: In some reforming processes, a degree of hydrocracking may occur. Hydrocracking involves the breaking of larger hydrocarbon molecules into smaller fragments. This can help convert heavier components into lighter, more valuable products such as gasoline.

Reforming is typically performed at elevated temperatures and pressures. There are two primary categories into which the process can be classified:

1. Continuous Catalytic Reforming (CCR): In continuous catalytic reforming, the feedstock is continuously passed through a series of reactors containing the catalyst. The reactions occur under elevated temperatures and moderate pressures. The catalyst is continuously regenerated to maintain its activity and ensure optimal performance.

2. Semi-Regenerative Reforming (SRR): In semi-regenerative reforming, the feedstock is passed through a fixed-bed reactor containing the catalyst. Once the catalyst becomes deactivated, it is removed from the reactor, regenerated, and then returned for reuse. The process operates in cycles, alternating between reforming and catalyst regeneration stages.

The main product of reforming is high-octane gasoline, which is an essential component of transportation fuels. The improved octane rating of the gasoline enhances its resistance to knocking and improves engine performance. Additionally, reforming can produce byproducts such as hydrogen gas, light gases, and aromatics, which have various applications in the refining and petrochemical industries. Reforming plays a crucial role in the petroleum refining process as it helps meet the demand for high-quality gasoline with improved performance characteristics. By converting low-octane naphtha into higher-octane products, reforming contributes to the production of cleaner-burning fuels and supports the efficient operation of internal combustion engines

TREATING AND PURIFICATION

Treating and purification are essential steps in the refining process that involve the removal of impurities and contaminants from crude oil and refined petroleum products. The goal of these processes is to improve the quality of the final products and to ensure that they meet industry and regulatory standards. There are several types of treating and purification processes used in petroleum refining, including:

1. Desulfurization: Desulfurization is a process used to remove sulfur compounds from crude oil and refined products. Sulfur compounds can be harmful to the environment and can damage engines and other equipment. Desulfurization can be achieved through a variety of methods, including hydrodesulfurization, which involves the use of hydrogen gas and a catalyst to break down sulfur compounds, and adsorption, which involves the use of materials that can selectively adsorb sulfur compounds.

2. Hydrotreating: Hydrotreating is a process used to remove impurities such as sulfur, nitrogen, and metals from crude oil and refined products. Hydrotreating involves the use of hydrogen gas and a catalyst to break down impurities into smaller, less harmful molecules.

3. Distillation: Distillation is a process used to separate and purify crude oil and refined products. Distillation involves the heating of the material to vaporize the components, which are then condensed and collected. The different boiling points of the components allow for their separation.

4. Filtration: Filtration is a process used to remove solid impurities from crude oil and refined products. Filters are used to trap solid particles and prevent them from entering downstream equipment or products.

5. Adsorption: Adsorption is a process used to remove impurities such as sulfur and nitrogen compounds from refined products. Adsorption involves the use of materials that can selectively adsorb the impurities, which are then removed from the product.

6. Decanting and centrifugation: Decanting and centrifugation are processes used to separate liquids and solids. These processes involve the use of gravity or centrifugal force to separate the components based on their density.

Treating and purification processes are important for ensuring the quality and purity of petroleum products. They help to meet industry and regulatory standards, and ensure that the products are safe for use in engines and other equipment. These processes also help to reduce the environmental impact of petroleum products by removing harmful impurities and contaminants.

BLENDING

Blending in the context of petroleum refining refers to the process of combining different refined petroleum products and additives to create final products with desired properties and specifications. Blending is a crucial step in the refining process as it allows for the customization of products to meet market demands and regulatory requirements.

The blending process comprises the subsequent stages:

1. Product Selection: Different refined petroleum products, such as gasoline, diesel, and jet fuel, are selected based on their specific characteristics, such as octane rating, cetane number, volatility, and other performance parameters. These products serve as the base components for blending.

2. Additive Addition: Various additives may be introduced to enhance the performance and quality of the final product. Additives can include detergents, anti-oxidants, lubricity agents, corrosion inhibitors, and antiknock agents, among others. These additives help improve fuel efficiency, reduce emissions, protect engine components, and meet regulatory requirements.

3. Mixing and Homogenization: The selected petroleum products and additives are mixed together to achieve a homogenous blend. This can be accomplished through mechanical mixing, such as with agitators or pumps, or through recirculation in blending tanks. The goal is to ensure that the components are thoroughly mixed to create a uniform blend.

4. Quality Control: Blended products undergo quality control testing to verify that they meet the desired specifications and regulatory standards. Various parameters, such as octane rating, flash point, viscosity, sulfur content, and density, are analyzed to ensure compliance.

5. Final Product Distribution: Once the blending process is complete and the product meets the required specifications, the final blend is typically stored in storage tanks or transported via pipelines or tanker trucks for distribution to end-users, such as gas stations or airports.

Blending allows refiners to produce a wide range of products with specific properties to meet market demand. It enables the production of different grades of gasoline, diesel, and other petroleum products that cater to specific requirements, such as different climate conditions or engine types.

Blending also provides flexibility in adjusting product characteristics to comply with regional fuel standards and regulations. Furthermore, blending is not limited to finished products. It can also be applied at various stages of the refining process to optimize intermediate product streams. For example, blending may be performed to modify the properties of naphtha or gas oil fractions to meet specific processing requirements or to create feedstock for downstream units. Overall, blending is a critical process in petroleum refining that enables refiners to create tailored products with specific properties and meet market demands. It ensures that the final products meet quality specifications, regulatory requirements, and the performance needs of end-users.

CHAPTER VIII. ENVIRONMENTAL CONSIDERATIONS

POLLUTION AND EMISSION CONTROL

Pollution and emission control in the petroleum refining industry are crucial for reducing the environmental impact of refining operations and meeting regulatory requirements. Refineries implement various measures to control and mitigate pollution and emissions throughout the refining process. Below are some key strategies and technologies used for pollution and emission control:

1. Process Optimization: Refineries employ process optimization techniques to maximize efficiency and minimize waste generation. This includes optimizing operating conditions, such as temperature and pressure, to improve process efficiency and reduce emissions.

2. Advanced Control Systems: Refineries use advanced control systems and automation to monitor and control operations, ensuring optimal performance and minimizing the potential for emissions.

3. Flue Gas Treatment: Flue gas treatment technologies, such as flue gas desulfurization (FGD) and selective catalytic reduction (SCR), are employed to remove pollutants from the flue gases emitted from combustion sources like boilers, heaters, and furnaces. FGD removes sulfur dioxide ($SO2$), while SCR reduces nitrogen oxide (NOx) emissions.

4. Low NOx Burners: Low NOx burners are used in combustion processes to minimize the formation of nitrogen oxides (NOx), which are a major contributor to air pollution and smog. These burners achieve efficient fuel combustion while reducing the release of harmful pollutants.

5. Wastewater Treatment: Refineries implement wastewater treatment systems to remove pollutants from process water and ensure compliance with regulatory discharge standards. Treatment methods include physical, chemical, and biological processes to remove contaminants before the water is discharged or reused.

6. Vapor Recovery Units (VRUs): Vapor recovery units' capture and recover volatile organic compounds (VOCs) released during various stages of the refining process, such as storage tanks, loading and unloading operations, and product transfers. These units prevent VOC emissions into the atmosphere and may recover valuable hydrocarbons for further use.

7. Effluent Treatment: Refineries treat process effluents, including spent caustics and other chemical waste streams, to remove harmful contaminants before disposal. Treatment processes can include neutralization, sedimentation, filtration, and biological treatment.

8. Continuous Monitoring: Refineries implement continuous monitoring systems to track emissions and pollutant levels in real-time. This allows for prompt detection of deviations from regulatory limits and enables timely corrective actions.

9. Renewable Energy Integration: Refineries are exploring the integration of renewable energy sources, such as solar or wind power, to supplement their energy needs. This reduces reliance on fossil fuels and lowers greenhouse gas emissions associated with energy generation.

10. Environmental Management Systems (EMS): Refineries implement EMS frameworks to systematically identify, monitor, and manage environmental aspects and impacts. EMS helps to ensure compliance with environmental regulations, establish improvement targets, and promote a culture of environmental stewardship.

These pollution and emission control measures help refineries minimize their environmental footprint, reduce air and water pollution, and mitigate the potential adverse effects of refining operations. By implementing advanced technologies and adopting sustainable practices, refineries aim to operate in an environmentally responsible manner while producing high-quality petroleum products.

ENERGY EFFICIENCY AND CONSERVATION

Energy efficiency and conservation play significant roles in the petroleum refining industry as they contribute to reducing energy consumption, lowering operating costs, and minimizing environmental impacts. Refineries implement various strategies and technologies to improve energy efficiency and conserve energy throughout their operations.

1. Process Optimization: Refineries continuously optimize their processes to maximize energy efficiency. This includes optimizing operating conditions, such as temperature and pressure, to minimize energy requirements while maintaining product quality. Advanced process control systems and modeling tools are used to identify opportunities for process optimization and energy savings.

2. Heat Integration: Refineries employ heat integration techniques to recover and reuse heat from different process streams. Heat exchangers are used to transfer heat between hot and cold streams, reducing the need for additional heating or cooling energy. This approach improves overall energy efficiency and reduces energy consumption.

3. Cogeneration and Combined Heat and Power (CHP): Refineries often generate their own electricity and heat through cogeneration or combined heat and power systems. These systems produce both electrical power and useful heat from a single fuel source, such as natural gas or biomass. The waste heat from power generation is recovered and used for various process heating requirements, further improving energy efficiency.

4. Energy Management Systems (EMS):
Refineries implement EMS frameworks to monitor, manage, and optimize energy consumption. Real-time energy monitoring, energy performance indicators, and data analytics help identify energy-saving opportunities, track energy usage, and facilitate informed decision-making for energy management.

5. Equipment Upgrades and Maintenance:
Refineries invest in upgrading and maintaining their equipment and infrastructure to improve energy efficiency. This can include retrofitting older equipment with more efficient models, optimizing insulation, and ensuring proper maintenance and calibration of equipment to minimize energy losses.

6. Energy Recovery Systems: Refineries employ energy recovery systems to capture and utilize waste energy. For example, waste heat recovery systems recover heat from flue gases, exhaust streams, and other waste streams for use in process heating or power generation, reducing the need for additional energy inputs.

7. Lighting and HVAC Optimization: Refineries optimize lighting systems and heating, ventilation, and air conditioning (HVAC) systems to reduce energy consumption. This includes using energy-efficient lighting fixtures, occupancy sensors, and programmable thermostats to optimize energy usage in non-process areas.

8. Employee Awareness and Training: Refineries promote energy conservation and efficiency through employee awareness programs and training. Engaging employees and encouraging energy-saving behaviors can lead to significant energy savings and foster a culture of energy efficiency throughout the organization.

9. Renewable Energy Integration: Refineries are exploring the integration of renewable energy sources, such as solar and wind power, into their operations. This reduces reliance on fossil fuels and contributes to a more sustainable energy mix.

CHAPTER IX. CHALLENGES AND FUTURE TRENDS

CHANGING CRUDE OIL QUALITIES

Changing crude oil qualities can have a significant impact on the refining process and the resulting petroleum products. Crude oil qualities can vary based on factors such as geographic location, source, and extraction methods. Changes in crude oil qualities can occur due to shifts in the global oil market, availability of specific crude grades, and refining strategies. Here are some important factors to consider and the impacts associated with changing crude oil qualities:

1. Feedstock Selection: Refineries may need to adapt their feedstock selection based on the changing qualities of available crude oils. Different crude oil qualities have varying compositions of hydrocarbons, sulfur content, viscosity, and other properties. Refineries select crude oils that are suitable for their processing capabilities and desired product slate.

2. Processing Capacity and Equipment:
Changing crude oil qualities may require adjustments or modifications to refinery processing units. Some crude oils may contain higher levels of impurities or require different processing conditions compared to others. Refineries need to ensure that their processing units are equipped to handle the specific properties and characteristics of the crude oil being processed.

3. Product Yield and Quality: Crude oil qualities can affect the yield and quality of the resulting petroleum products. Different crude oils have varying proportions of light, middle, and heavy hydrocarbons. This can impact the yield of gasoline, diesel, jet fuel, and other product fractions. Additionally, the presence of impurities like sulfur and metals in crude oils can influence the quality and specifications of the final products.

4. Refinery Configuration: Refineries may need to optimize or adjust their configuration to handle changing crude oil qualities effectively. This can involve reconfiguring units, adding new equipment, or upgrading existing infrastructure to accommodate specific crude oil characteristics.

5. Process Flexibility: Refineries with greater process flexibility can better handle changing crude oil qualities. Flexibility refers to the refinery's ability to process a wide range of feedstocks and adjust operating conditions to optimize yields and product quality. Refineries with more versatile processing units can adapt to varying crude oil qualities without significant disruptions.

6. Blending and Product Adjustments: Refineries often employ blending techniques to manage the variations in crude oil qualities. Blending different crude oils with complementary properties can help achieve desired product specifications. Additionally, adjustments can be made to refining processes, such as catalyst selection or operating parameters, to optimize product quality when dealing with different crude oil qualities.

7. Economic Considerations: Changing crude oil qualities can impact refinery economics. Some crude oils may be more expensive to process due to higher impurity content or complex refining requirements. Refineries need to assess the economic viability and profitability of processing different crude oil qualities based on factors such as market demand, refining margins, and investment costs.

8. Environmental Considerations: Crude oil qualities can influence the environmental impact of refining operations. Crudes with higher sulfur content or heavier fractions may result in higher emissions and require additional pollution control measures. Refineries need to ensure compliance with environmental regulations and implement appropriate measures to minimize the environmental impact associated with processing specific crude oil qualities.

DEVELOPMENT OF ALTERNATIVE ENERGY SOURCES

The development of alternative energy sources is an important focus worldwide due to the need to reduce greenhouse gas emissions, address climate change, and diversify energy supplies. Alternative energy sources offer cleaner and more sustainable alternatives to traditional fossil fuels. Here are some significant advancements in the realm of alternative energy sources:

1. Renewable Energy: Renewable energy sources, such as solar, wind, hydroelectric, geothermal, and biomass, have witnessed significant development and deployment in recent years. These sources harness natural processes or resources that are continuously replenished, making them sustainable and environmentally friendly. Advances in renewable energy technologies, cost reductions, and policy support have facilitated their rapid growth and increased their share in the global energy mix.

2. Solar Power: Solar power has seen remarkable progress with the development of more efficient photovoltaic (PV) technologies and increased manufacturing scale. The cost of solar panels has significantly decreased, making solar energy more economically competitive. Large-scale solar power plants and distributed solar installations, such as rooftop solar panels, are being deployed worldwide to generate electricity.

3. Wind Power: Wind power has become one of the fastest-growing sources of renewable energy. Advances in wind turbine technology, including larger and more efficient turbines, have improved the cost-effectiveness and energy production of wind farms. Offshore wind farms are being developed in coastal areas, taking advantage of strong and consistent winds.

4. Hydropower: Hydropower has long been a major source of renewable energy, utilizing the energy of flowing or falling water to generate electricity. The development of more efficient turbines and improved environmental impact assessment and mitigation techniques have supported the growth of hydropower projects, including large-scale dams and smaller run-of-river installations.

5. Geothermal Energy: Geothermal energy taps into the heat generated by the Earth's internal processes. It involves extracting heat from underground sources to produce electricity or provide heating and cooling for buildings. Ongoing technological advancements and exploration of geothermal resources have expanded the utilization of this clean and reliable energy source.

6. Biomass Energy: Biomass energy utilizes organic materials, such as agricultural residues, forest waste, and dedicated energy crops, to produce heat, electricity, or biofuels. The development of advanced biomass conversion technologies, such as biomass gasification and bio-refineries, has enhanced the efficiency and sustainability of biomass energy production.

7. Energy Storage: Energy storage technologies, such as batteries, pumped hydro storage, and thermal energy storage, are crucial for the integration of renewable energy sources into the grid. Advances in energy storage systems enable the capture and efficient utilization of intermittent energy from sources like solar and wind, enhancing grid stability and enabling better integration of renewables.

8. Hydrogen and Fuel Cells: Hydrogen is emerging as a promising alternative energy carrier. It can be produced from renewable sources through electrolysis and used for various applications, including transportation and electricity generation through fuel cells. The development of cost-effective hydrogen production methods and fuel cell technologies is advancing the use of hydrogen as a clean energy source.

9. Carbon Capture and Utilization (CCU) and Storage (CCS): CCU and CCS technologies aim to capture and store or utilize carbon dioxide (CO_2) emissions from power plants and industrial facilities. These technologies help reduce greenhouse gas emissions from fossil fuel-based energy sources and mitigate climate change.

10. Energy Efficiency and Conservation: Alongside the development of alternative energy sources, improving energy efficiency and conservation practices is crucial. Efforts to reduce energy waste, promote energy-efficient technologies, and enhance energy management systems can significantly contribute to achieving sustainability goals.

The development of alternative energy sources is driven by a combination of technological advancements, policy support, and growing public awareness of the need for cleaner and more sustainable energy solutions. Continued research, investment, and collaboration between governments, industries, and academia are vital for further advancements and widespread adoption

ADVANCEMENTS IN REFINING TECHNOLOGIES

Advancements in refining technologies have been instrumental in improving the efficiency, flexibility, and environmental performance of petroleum refining processes. These advancements aim to optimize the conversion of crude oil into valuable petroleum products while reducing energy consumption, emissions, and the environmental footprint of the refining industry.

Here are a few notable advancements in refining technologies:

1. Advanced Catalytic Processes: Catalytic processes play a vital role in refining, and advancements in catalyst development have led to more efficient and selective conversion reactions. For example, fluid catalytic cracking (FCC) units, which convert heavy hydrocarbons into lighter products, have seen improvements in catalyst design and regeneration techniques, resulting in higher conversion rates and improved product yields.

2. Hydroprocessing Technologies: Hydroprocessing technologies, including hydrotreating and hydrocracking, have undergone significant advancements. These processes involve the use of hydrogen to remove impurities, such as sulfur and nitrogen, from crude oil and convert heavy fractions into lighter, high-quality products. Advanced catalysts and reactor designs have improved the efficiency and selectivity of hydroprocessing units.

3. Residue Upgrading: Residue upgrading technologies focus on converting heavy residues, such as vacuum residue, into more valuable products. Various processes, including delayed coking, solvent deasphalting, and visbreaking, have been enhanced to maximize the conversion of heavy residues into lighter fractions, such as gasoline, diesel, and jet fuel.

4. Integrated Refining and Petrochemical Processes: Integration between refining and petrochemical processes has gained attention in recent years. By incorporating petrochemical production units within the refinery complex, the production of high-value chemicals and polymers can be optimized, leading to increased profitability and reduced energy consumption.

5. Process Intensification: Process intensification aims to enhance the efficiency and productivity of refining processes by reducing equipment size, improving heat and mass transfer, and optimizing process conditions. Technologies such as microreactors, microchannel heat exchangers, and membrane separations are being explored to intensify key refining processes, resulting in higher yields, energy savings, and smaller footprints.

6. Advanced Separation Technologies: Refinery separation units, such as distillation columns and extractors, have seen advancements in design and operation. Improved internals, enhanced control strategies, and optimized column configurations have led to better separation efficiency, reduced energy requirements, and increased product yields.

7. Digitalization and Data Analytics: The application of digital technologies, such as artificial intelligence (AI), machine learning, and data analytics, is transforming the refining industry. These technologies enable real-time monitoring and optimization of refinery operations, predictive maintenance, and the identification of process bottlenecks and opportunities for improvement. Digital twins, which are virtual replicas of physical refining units, allow for advanced simulations and optimization of refinery processes.

8. Environmental and Energy Management Systems: Refineries are implementing advanced environmental and energy management systems to monitor and optimize energy consumption, emissions, and environmental performance. These systems integrate real-time data monitoring, modeling, and optimization algorithms to achieve energy efficiency targets, reduce emissions, and ensure compliance with environmental regulations.

9. Renewable Fuels and Biochemicals: Refineries are exploring the production of renewable fuels, such as biofuels and bio-based chemicals, to diversify their product portfolio and reduce dependence on fossil fuels. Advanced biofuels, produced from non-food biomass feedstocks, are being developed and integrated into existing refining processes.

10. Carbon Capture and Utilization (CCU) and Storage (CCS): Refineries are exploring carbon capture technologies to capture and store or utilize CO_2 emissions generated during the refining process. CCS and CCU technologies help reduce greenhouse gas emissions and mitigate climate change.

www.ingramcontent.com/pod-product-compliance
Lightning Source LLC
Chambersburg PA
CBHW060847220526
45466CB00003B/1273